Mathematica Successūs

Mikołaj Mocek

First Edition (last revision January 20, 2026)

Published by Mikołaj Mocek

Mathematica Successūs: A Formal Approach on Success, Systems and Self

For more information, visit “About the Author” page or contact at:
`mikolaj.mocek@icloud.com`

Goal of This Book

This is a short book for people who can read mathematics but are not yet where they want to be in life. Basically, it is a book written for my past self.

The aim is very specific:

- to describe *success* in any domain as a system that can be modeled in the language of sets, functions, probability and control;
- to extract from this modeling a few *hard constraints*—laws of success—that are as unforgiving as the laws of logic;
- to show how these laws connect to my own life: to things that failed, things that worked by accident, and things that started to work only when I saw them as mathematical objects.

The goal is not to sell hope. The self–help industry has a financial incentive to promise that "anyone can make it" if they buy just one more course. This book has a different promise:

> If you can follow definitions and proofs, you can get a toolbox for steering your life that is precise, brutally honest and reusable in any domain.

These tools will not remove luck, inequality or politics. They will not make institutions act in your interest. They will, however, help you see clearly *which parts of your success are under your control, which parts are structural, and where pushing harder is mathematically pointless.*

Structure. The book has five main chapters:

1. **Being** – what actually is; states, self and essential variables.

2. **Space** – what could be; possibility spaces, information and uncertainty.

3. **Change** – how states move; dynamics, feedback and learning.

4. **Goal** – what we call "success"; utility, decision and strategy.

5. **Application** – how to use the framework; a worked example of learning an ultra—hard skill.

Each part mixes formal definitions, small theorems and short notes from my life. Sometimes the math will look much more serious than the examples. That is on purpose: I want the theorems to stay true even when my own stories are wrong, biased or incomplete.

Contents

1 Being: States, Self and Essential Variables **1**

1.1 World as a Set of States 1
1.2 The Self as a Subsystem 2
1.3 Variables and Observables 3
1.4 Essential Variables and Survival 4
1.5 Success as a Property of States 5

2 Space: Possibility, Variety and Uncertainty **6**

2.1 Reachable and Feasible Sets 6
2.2 Variety and Information 7
2.3 Only Variety Can Destroy Variety 8
2.4 Uncertainty and Belief: Probability as Logic 9
2.5 Information as a Resource 9

3 Change: Dynamics, Feedback and Learning **11**

3.1 Dynamics and Control 11
3.2 Feedback Loops 12
3.3 Law of Requisite Variety In Regulation 12
3.4 Learning as Bayesian Feedback 13
3.5 Ultrastability and Meta-Systems 16

4 Goal: Utility, Decision and Strategy **17**

4.1 Utility Functions and Multi-Dimensional Success . . . 17
4.2 Decision Theory and Utility Maximization 19
4.3 Planning as Control Under Uncertainty 20
4.4 Macro-Levers, Micro-Levers and Daily Systems 21
4.5 The Limits of Control 22

5 Application: Learning an Ultra—Hard Skill 24
5.1 Define the Model: States, Self, Essential Variables . . . 24
5.2 Reachable Set and Requisite Variety 25
5.3 Policy Design: Micro—Levers for a Hard Skill 25
5.4 Policy and Bayesian Learning Loop 26
5.5 Failure Modes and Replans 26
5.6 Complementary Completion and Transfer 27

Chapter 1

Being: States, Self and Essential Variables

This chapter answers the first question: *what actually is*? Before we can optimize, we must be clear about what exists in our model of the world. We will build a vocabulary of states, selves, and essential variables. These definitions form the foundation for everything that follows: they allow us to speak about change (Chapter 3), possibility (Chapter 2), and finally success (Chapter 4).

1.1 World as a Set of States

We start from the dullest possible place: a set.

Definition 1.1 (World). A *world* is a nonempty set X. An element $x \in X$ is called a *state*.

At first, this looks too abstract. But almost every serious textbook you know starts this way: set theory, analysis, probability, control.[1] Why set theory? Because sets are the simplest structure that lets us talk about "things" and "collections of things" without any additional assumptions. Once we have sets, we can define functions, relations, and eventually all the machinery of probability and control. Starting anywhere else would smuggle in hidden assumptions.

In this book, X is "whatever matters for the success story we care about", it is the "everything". It could be:

[1]For example, a classic set theory text builds *everything* from sets and membership relations.

- your health, money, social ties and knowledge;
- the state of a company (cash, product quality, relationships);
- the state of a market, a political system or an ecosystem.

First challenge. In real life X is enormous and mostly unknown. You never see the full state. So the first move in any serious attempt at success is already a lie: you choose some simplified X and pretend that it is "the world".

Remark 1.1 (Modeling honesty). You will never get away from this. The trick is to be explicit:

> "This is my model X. It forgets many things. I am using it because it is simple enough that I can compute decisions. I accept that using this model means I might win locally and lose globally."

1.2 The Self as a Subsystem

In most self–help books the author talks about "you" as if it was obvious what that word means. Here we will be slightly more precise.

Definition 1.2 (Self). Let X be a world. The *self* is a distinguished subsystem

$$S \subseteq X$$

together with some variables we can influence, called *controls*.

In practice, S includes your body, your brain and the artifacts you can reliably use (phone, notebook, routines, money).

When I was younger, my mental model was different. I felt like a point moving through time, and everything outside my skin was "the world". Later, when I read about systems and control, I realised that this view was too small. A well–designed routine or a computer script is as much "me" as my hand is. It is a piece of my extended self, because it moves when I choose and keeps moving even when I forget.

1.3 Variables and Observables

We do not see states directly. We see measurements. This is not a philosophical curiosity but a hard constraint: as finite beings, we cannot observe all states the world presents us. We must map them to something we can grasp—numbers, categories, symbols. This reduction is what observables do.

Definition 1.3 (Observable)**.** Let X be a world. An *observable* is a function

$$Y : X \to V$$

from states to some value set V (numbers, categories, vectors, …).

Examples:

- bank balance $B : X \to \mathbb{R}$ (in theory),
- body weight $W : X \to \mathbb{R}$ (in theory),
- number of close friends $F : X \to \{0, 1, 2, \dots\}$,
- subjective mood $M : X \to \{-2, -1, 0, 1, 2\}$.

You only get to interact with the world through a finite list of such observables and a finite list of actions. Already here a first law of success appears.

Principle 1.1 (Representation precedes optimization)**.** You can only optimize what you can represent as an observable.

If something important is not tracked anywhere—not in your notebook, not in numbers, not even in a short phrase—then in the mathematics of your life it does not exist. You will sacrifice it accidentally.

Remark 1.2 (The cost of implicit variables)**.** If you optimize only for what you measure and ignore what you do not, you will inadvertently degrade the unmeasured dimensions. The fix is simple but requires discipline: identify what matters to you, make it observable, and include it in your objective. Once you track sleep alongside productivity, or relationships alongside achievement, your behavior shifts naturally because the optimization surface itself has changed.

1.4 Essential Variables and Survival

There are variables that *must* stay in a certain range if we want to survive and survival is the foundation of any success.

We should make a sharp distinction between variables that *must* stay in a certain range and variables that can swing freely.[2]

The essential variables are the super set that limits us from below. No matter how much money we make, if our health markers go out of range, we fail. Later in this book we will introduce reachable 2.1 sets, the essential variables are the largest reachable set that still allows survival.

Definition 1.4 (Essential variables). Let $E = (E_1, \dots, E_k)$ be a list of observables $E_i : X \to \mathbb{R}$. A subset $R \subset \mathbb{R}^k$ is called the *survival region.* The system *survives* as long as

$$E(x) \in R.$$

In biology, the E_i are things like blood pressure, temperature, oxygen level. The survival region R is the range in which the organism can stay alive. The world throws disturbances at the organism; the organism acts to pull E back into R.

This idea scales to life design:

- health markers (sleep, pain, energy),
- finances (cash buffer, debt level),
- relationships (number of people you can ask for help),
- integrity (how often you lie to yourself or others).

Remark 1.3 (Life–level survival). You can define your own vector of "essential variables". Then you can declare:

> "I am *not* successful if any of these leaves its survival region, even if I make a lot of money or get a lot of attention."

This sounds moral, but it is also mathematical. It defines a region in state space where "failure" is guaranteed no matter what other numbers say.

A very brutal lesson of systems theory is:

[2]The classic reference is Ashby's work on essential variables and regulation.

Principle 1.2 (First law: stability before optimisation). If your essential variables are not stable, optimising secondary variables is mathematically meaningless.

If your cash buffer is near zero, your body is falling apart and your relationships are collapsing, then trying to "maximise learning" or "maximise impact" is like maximising a function outside its domain.

I did exactly this at some point. I tried to do hard academic-level math and business while sleeping little, carrying chronic stress and ignoring my bank account. From the outside it looked heroic. From the inside it was just a system violating its own survival constraints, slowly but systematically melting down.

1.5 Success as a Property of States

We can now speak about success without poetry.

Definition 1.5 (Success predicate). A *success predicate* on X is a function

$$S : X \to \{0, 1\}$$

that is 1 for states we call "successful" and 0 otherwise.

The important thing is that S need not be global. It is almost always *personal and local.* For one person, "success" means having a stable, low–stress job and a warm family. For another, it means high risk, high status, public impact.

Remark 1.4 (No universal success set). There is no single subset $S^{-1}(1) \subset X$ where everyone agrees that those states are "true success". Any book that talks as if there was just one is either selling ideology or a product.

What we *can* do, objectively, is to show constraints that any S must obey if you want it to be reachable from where you are, given the structure of the world and your own limits. That is what the next chapters do.

Chapter 2

Space: Possibility, Variety and Uncertainty

This chapter answers the second question: *what could be*? Using the vocabulary from Chapter 1, we now map the landscape of possibility. We define which states are reachable from where we are, how much variety exists in both the world and our own actions, and how to reason about uncertainty using probability. These tools allow us to see the constraints on what success can mean for us. Later, in Chapter 3, we will use this map to understand how systems move. In Chapter 4, we will use it to choose our goals.

2.1 Reachable and Feasible Sets

Not every state in X is reachable from where you are. Some are blocked by physics. Others by institutions, by other agents, or just by time.

Definition 2.1 (Reachable set). Let $x_0 \in X$ be the current state. A subset $R(x_0) \subseteq X$ is called the *reachable set* if it contains all states that can be reached from x_0 by some sequence of allowed actions within some time horizon.

We will formalise actions and dynamics in the next chapter. For now, just remember that any realistic success plan lives inside $R(x_0)$.

Remark 2.1 (Structural injustice as a constraint). Some books talk about success as if any state was reachable if you just "believed" hard enough. In reality, $R(x_0)$ is heavily shaped by things you did not choose: which country you were born in, which passports you have, who your parents are, how people perceive your race and gender, and what institutions want from you.

Pretending that $R(x_0)$ is the same for everyone is not optimism; it is a deliberate distortion introduced at the modeling stage.

2.2 Variety and Information

Variety represents the amount of choice or uncertainty in a set and is a tool that will allow us to quantify complexity. Why do we need this? Because success requires navigating a world with many possible states, and the number of states we can distinguish or respond to determines what we can achieve. Variety is the bridge between abstract set theory and practical control.

We use the word *variety* to measure how many distinct states or messages a system can take.[1] [2]

Definition 2.2 (Variety). Let A be a finite set with N elements. The *variety* of A is

$$V(A) = \log_2 N.$$

If A is the set of all possible disturbance patterns, $V(A)$ tells us how much information there is in knowing which disturbance occurred. If A is the set of actions available to you, $V(A)$ tells us how much choice you have.

Example 2.1 (Simple career variety). If you only consider three options after graduation (corporate job, academia, or starting a company), then at that granularity the variety is $\log_2 3 \approx 1.585$ bits. If you know in advance that one option is impossible (say, visas), the variety drops.

This may look toy–like, but when we later write inequalities on varieties, these numbers become hard constraints.

Remark 2.2 (Variety and the big picture). Variety connects a lot in this book. Your reachable set $R(x_0)$ has some variety (how many distinct futures are accessible). Disturbances from the environment have variety

[1] Variety is usually defined as the logarithm of the number of distinguishable states, connecting it to information and entropy.

[2] A bit is the amount of information gained when the uncertainty about a choice is reduced by half. It represents a single binary decision—a *Yes/No* or *0/1* choice which allows us to chain it and make decision chains that discretely represent the reality.

V_D. Your action repertoire has variety V_R. The law of requisite variety then tells you: if $V_R < V_D$, you cannot fully regulate the outcome. This is why expanding skills, tools, and strategies matters—they increase V_R. And this is why simplifying your environment or specializing can help—they decrease V_D. Variety is one the most significant elements in understanding success.

2.3 Only Variety Can Destroy Variety

The *law of requisite variety* is one of the few almost–theorems about control and success. Informally it says:

> To reliably keep some important variables in a safe range in the presence of disturbances with variety V_D, your regulator (skills, tools, systems) must have at least comparable variety V_R in its repertoire of actions. Only variety can destroy variety.

Formally, in one of its simplest forms, it can be written as an inequality

$$V_O \geq V_D - V_R,$$

where V_O is the remaining variety in the outcome after regulation.

Remark 2.3 (Interpreting the inequality). If disturbances are complex (large V_D) and your reaction repertoire has small variety (small V_R), then V_O cannot be small. You will see chaos in your results no matter how hard you "try".

There are only two ways out:

1. increase V_R: learn more skills, use better tools, delegate to more diverse agents, build richer strategies;
2. reduce V_D: change environment, specialise, impose constraints that cut off many kinds of disturbance.

This applies to any domain: social, professional, or personal. A system cannot regulate complexity it has no mechanisms to sense or influence.

2.4 Uncertainty and Belief: Probability as Logic

So far we treated sets as if we knew exactly which state we are in. In real life we almost never do. We guess.

Probability theory, in the Bayesian view, is not about "randomness out there" but about degrees of belief inside our head.[3]

Definition 2.3 (Belief state). A *belief state* about the world X is a probability measure P on X assigning to each event $A \subseteq X$ a number $P(A) \in [0, 1]$ obeying the usual axioms.

Principle 2.1 (Coherence law). If your degrees of belief violate the probability axioms, there exists a bet structure that guarantees you a loss (a Dutch book), no matter what happens. Therefore, *coherent* success planning requires that beliefs obey these axioms.

Remark 2.4 (What is a Dutch book?). A Dutch book is a set of bets that, taken together, guarantee a loss regardless of the outcome. For example, if you believe event A has probability 0.6 and also believe $\neg A$ has probability 0.6, a clever bettor can construct wagers that extract money from you no matter what happens. The name comes from historical gambling terminology. The theorem states: if your beliefs are incoherent (violate probability axioms), such a losing bet structure always exists.

Remark 2.5 (Average vs optimal). Here we see a first separation between "average" and "optimal". Many people live with incoherent beliefs: they over– or under–estimate rare events, they believe mutually inconsistent stories. They might still do okay, because the world is forgiving. But if you want a strategy that is provably non–self–defeating, you cannot ignore coherence. It is like using arithmetic without respecting $2 + 2 = 4$.

2.5 Information as a Resource

If beliefs are probabilities, *information* is whatever changes those probabilities.

Formally, observing new data D updates P into a posterior $P(\cdot|D)$ via Bayes' rule. The notation $P(\cdot|D)$ means "the probability distribution

[3]This viewpoint is developed in detail by Jaynes (see bibliography), who treats probability as extended logic for reasoning with incomplete information.

after conditioning on having observed D". The dot $\cdot$ is a placeholder for whatever event or state we want to query.[4]

In success terms:

- information lets you cut down the space of possible worlds,
- which lets you choose better actions,
- which lets you achieve success with less wasted effort.

This sounds clean, but there is a complication: information is not free, and information channels are not neutral.

Remark 2.6 (Incentives and censorship)**.** News feeds, social platforms, school curricula and corporate dashboards are not designed to maximise your personal expected utility. They are optimised for the goals of their creators: attention, ad revenue, political stability, brand reputation.

If you treat all incoming information as unbiased and complete, your belief state will be systematically distorted. This is not a conspiracy theory, just a basic fact about incentives.

A mathematically minded way to defend yourself is to think of each source as a noisy channel with a known or guessed bias, and to adjust your update rules accordingly. We will not go deep into this, but the main point stands: information is a resource, and not all information streams are on your side.

---

[4]Bayes' rule and maximum-entropy methods give a principled way to update beliefs; their justification as uniquely consistent rules is a key point in modern Bayesian theory.

Chapter 3

Change: Dynamics, Feedback and Learning

This chapter answers the third question: *how do states move*? Given the world (Chapter 1) and its geometry of possibility (Chapter 2), we now ask: by what rules do systems evolve over time? We define dynamics, policies and feedback loops. We show how the law of requisite variety appears again, this time as a constraint on how complex a system can be if we want to stabilize it. We show how learning works as Bayesian feedback: updating our model of the world based on what we observe. These dynamics form the machinery that connects where we are to where we want to go.

3.1 Dynamics and Control

Definition 3.1 (Dynamics with control). Let X be the state space, U the action space and W the disturbance space. A *controlled dynamic system* is given by an update rule

$$x_{t+1} = f(x_t, u_t, w_t),$$

where $x_t \in X$ is the state, $u_t \in U$ the control action, and $w_t \in W$ the disturbance at time t.

To actually use this, we often represent f as a feedback diagram with blocks and arrows, like in a control textbook.[1]

[1]Standard feedback–systems texts present many such diagrams and analyse stability and performance of the resulting closed–loop system.

In life, f is unknown, time–varying and partly adversarial. But we can still design *policies.*

Definition 3.2 (Policy). A *policy* is a function π that selects an action u_t based on the history of observations:

$$u_t = \pi(y_0, \ldots, y_t),$$

where y_t are observations produced from x_t.

Your habits are policies. Your study schedule is a policy. Your reaction to criticism is a policy. Most of them are implicit until you write them down.

3.2 Feedback Loops

Definition 3.3 (Feedback). A system uses *feedback* if its current action depends on its own past outputs or states.

Feedback is powerful and dangerous: it can stabilise a system (like a thermostat or a PID controller), or it can generate oscillations, chaos or blow–up.

Remark 3.1 (Self–reinforcing loops). When a feedback loop amplifies perturbations instead of damping them, small disturbances grow. For example, a social situation where nervousness triggers awkward behavior, which triggers more rejection, which triggers more nervousness, creates a destabilizing cycle. No amount of resolve can fix this until the loop structure itself changes. Control theory explains this not as a character flaw but as a mathematical property of the map f and feedback law π.[2]

3.3 Law of Requisite Variety In Regulation

We now embed the law of requisite variety into dynamics.

Let E_t be the vector of essential variables at time t, and let R be our survival region. Disturbances w_t push E_t out of R; the regulator uses actions u_t to keep E_t in R.

[2] In control theory, this corresponds to an unstable closed-loop system where the loop gain exceeds unity for the relevant frequency components. The fix is to change the feedback law π (your response pattern) or the plant f (the environment's dynamics).

The law of requisite variety tells us that if the disturbance process has variety V_D and the regulator's action set has variety V_R, then even with optimal policy the residual variety in E_t cannot go below $V_D - V_R$.

The precise relationship between varieties may involve scaling factors, measurement conventions, or irreducible time delays inherent in the dynamic system, which do not change the fundamental principle that V_O is bounded by $V_D - V_R$.

Principle 3.1 (Second law: requisite variety). Whenever you feel like life is "too complex" to manage, you can check the ratio between V_D and V_R. If you face many independent types of shock and you respond with only one or two fixed moves, instability is not a moral failure. It is linear algebra in action.

This law also has a dark side. It puts a ceiling on what any regulator can achieve, including institutions and leaders. A single decision-maker with finite bandwidth cannot stably control a complex society beyond a certain point, no matter how smart or good-willed they are.

3.4 Learning as Bayesian Feedback

We now formalise learning with standard Bayesian notation.[3]

Let Θ be a (possibly finite or continuous) parameter space indexing a family of models of the world. For each $\theta \in \Theta$ let

$$f_\theta : X \times U \times W \to X$$

be the state transition (or, more generally, a probabilistic transition kernel) and let

$$p_\theta(y_t \mid x_t, a_t)$$

be the likelihood of an observation y_t given current state x_t and action a_t under model θ.

Remark 3.2 (Proportionality). When you see the expression:

$$A \propto B$$

It means that A is directly proportional to B.

[3]Why Bayesian? Because Bayes' rule is the *unique* way to update beliefs that respects basic coherence requirements. Any other update rule either violates probability axioms or can be shown to be equivalent to Bayes under some transformation. If you want to learn from data without being Dutch-booked, you must update like a Bayesian.

Remark 3.3 (Sum). When you see $\sum$, it means that we are adding a series of terms together. For example:

$$\sum_{i=1}^{n} a_i = a_1 + a_2 + a_3 + \cdots + a_n$$

Remark 3.4 (Integral). When you see $\int$, it means that we are calculating the area under a curve. For example:

$$\int_a^b f(x)\,dx$$

represents the area under the curve of $f(x)$ from $x = a$ to $x = b$.

In the context of probability, integrals are often used to calculate expected values or probabilities over continuous random variables. In plain words it means "adding up" over a continuous range.

Definition 3.4 (Prior and posterior). A Bayesian learner starts with a prior probability measure P_0 on Θ. Given history

$$H_t = (y_0, a_0, \ldots, y_{t-1}, a_{t-1}, y_t),$$

the posterior P_t is defined by Bayes' rule:

$$P_t(d\theta) \propto L(H_t \mid \theta)\, P_0(d\theta),$$

where $L(H_t \mid \theta) = \prod_{s=0}^{t} p_\theta(y_s \mid x_s, a_s)$ is the likelihood of the history under θ. Equivalently, the recursive update is

$$P_{t+1}(d\theta) \propto p_\theta(y_{t+1} \mid x_{t+1}, a_{t+1})\, P_t(d\theta).$$

The posterior induces a predictive distribution for the next state/observation:

$$p(y_{t+1} \mid H_t, a_{t+1}) = \underbrace{\int_\Theta}_{\text{Average on all models } \theta} p_\theta(y_{t+1} \mid x_{t+1}, a_{t+1})\, P_t(d\theta),$$

and, if transitions are deterministic given θ, for the next state

$$p(x_{t+1}) = \underbrace{\int_\Theta}_{\substack{\text{Average on} \\ \text{all models } \theta}} \left(\underbrace{\int_W \mathbf{1}\{x_{t+1} = f(\ldots, w)\}\, p(dw)}_{\substack{\text{Probability of} \\ \text{the model } \theta}} \right) \underbrace{P_t(d\theta)}_{\substack{\text{Weight of} \\ \text{the model}}}$$

or the analogous kernel for stochastic transitions.

Definition 3.5 (Bayesian decision rule / learner). Given a utility function $U : X \to \mathbb{R}$, an action space A, and (optionally) a planning horizon H, the Bayesian learner selects actions by maximising posterior expected utility.

For a one-step myopic rule:

$$\pi_t(H_t) \in \underbrace{\arg\max_{a \in A}}_{\text{Select best action}} \quad \underbrace{\mathbb{E}_{\theta \sim P_t}}_{\substack{\text{Average over models } \theta \\ \text{weighted by belief}}} \left[\underbrace{\mathbb{E}_{x_{t+1} \mid H_t, a, \theta}\left[U(x_{t+1})\right]}_{\substack{\text{Expected payoff given} \\ \text{a specific model } \theta}} \right].$$

For a horizon H, we optimize the whole policy:

$$\pi_t \in \underbrace{\arg\max_{\pi}}_{\text{Select best rule}} \; \mathbb{E}_{\theta \sim P_t}\left[\mathbb{E}\left[\underbrace{\sum_{k=1}^{H} \gamma^{k-1} U(x_{t+k})}_{\substack{\text{Discounted sum of} \\ \text{future rewards}}} \;\middle|\; H_t, \pi, \theta \right]\right],$$

with discount factor $\gamma \in (0, 1]$, where discount factor is a number between 0 and 1 that determines how much future rewards are valued compared to immediate rewards.

Bayesian learning naturally balances exploitation and (posterior) exploration: actions that are informative about θ enter the inner expectation via their effect on future posteriors and hence on future utilities. One can make this explicit by adding an information gain term (e.g. expected reduction in entropy of P_t) to the objective when designing exploration strategies.

We can justify Bayes' rule as the consistent rational update for P_t under weak desiderata of coherence.[4]

Remark 3.5 (Model revision). If the posterior concentrates on models that still poorly predict observations, perform model revision: enlarge or change the model class Θ, or introduce latent structure. Formally, replace Θ by a richer family $\widetilde{\Theta}$ and initialise a prior $\widetilde{P}_0$ consistent with the new hypotheses; continue updates on the extended space.

[4]Jaynes, E. T. (2003). *Probability Theory: The Logic of Science.* Cambridge University Press.

3.5 Ultrastability and Meta–Systems

Sometimes both the environment and the regulator are so complex that we need a second level: a system that changes the way the first–level system changes.

We call this *ultrastability*:[5] a regulator that can change its own structure when it fails, searching over possible configurations until it finds one that keeps the essential variables in range.

Definition 3.6 (Ultrastable system). Let Π be a space of possible policies. An *ultrastable system* consists of:

1. a first-level regulator $\pi \in \Pi$ that acts on the state x_t;
2. a second-level meta-regulator $\mu : \Pi \times \text{History} \to \Pi$ that monitors whether essential variables $E(x_t)$ stay in the survival region R;
3. a switching rule: if $E(x_t) \notin R$ for sufficiently long, the meta-regulator selects a new policy $\pi' = \mu(\pi, H_t)$.

The system is ultrastable if, under mild conditions on the environment, the meta-regulator eventually finds a policy π^* such that $E(x_t) \in R$ for all future t.

In personal terms:

- first level (π): habits, schedules, scripts;
- second level (μ): a regular review process that examines which habits work, which ones are killing you, and replaces them.

When people say "I reinvented myself", often what actually happened is an ultrastable process: repeated crashes followed by structural changes until a stable configuration was found.

[5]For a classic reference, see Conant and Ashby's work on ultrastability.

Chapter 4

Goal: Utility, Decision and Strategy

This chapter answers the final question: *what do we call success*? We now bring together all the pieces: the states and observables (Chapter 1), the map of possibilities (Chapter 2), and the rules of change (Chapter 3). We define utility functions that encode our preferences, and we show how to make decisions under uncertainty that are provably coherent with those preferences. We close by showing how the framework applies to real life through micro-actions.

4.1 Utility Functions and Multi-Dimensional Success

A success predicate $S : X \to \{0, 1\}$ is too coarse—it only says “success” or “not success” without distinguishing degrees. We need to be able to talk about “better” and “worse” states, about trade-offs and risks. Later in Chapter 5, we will see how this generalization becomes essential for practical decision-making.

Definition 4.1 (Utility function). A *utility function* is a map

$$U : X \to \mathbb{R}$$

that encodes the relative preference of states: higher $U(x)$ means a more preferred state.

Utility functions are not unique; many different U can encode the same ordering of preferences. But the central idea is that under uncertainty, a rational agent chooses actions to maximise *expected utility.*[1]

Principle 4.1 (Third law: expected utility). Given a belief state P and a utility function U, any decision rule that systematically chooses actions with lower expected utility than some available alternative is dominated. Over time, it guarantees worse outcomes in expectation.

Multi–criteria utility

Real success is multi–dimensional: health, money, relationships, meaning, freedom, aesthetic pleasure, and more. There are several ways to handle this:

- scalarisation: build a weighted sum $U(x) = \sum_i w_i U_i(x)$;
- lexicographic order: some criteria are hard constraints (essential variables), others are optimised within the feasible region;
- vector utilities and Pareto optimality.

I personally use a hybrid:

1. Essential variables (sleep, basic health, not lying to myself, having at least two people I can call at 2am) are treated as survival constraints.
2. Within that region, I maximise a weighted sum of contribution, learning and autonomy.

This is not "the" correct U. It is just one explicit choice. The important thing is not which one you pick, but that you see clearly that you *are* picking something, and that many institutions around you will happily impose their own U if you do not.

[1] The standard expected–utility framework is also the basis of formal decision theory, including modern Bayesian treatments.

4.2 Decision Theory and Utility Maximization

Jaynes[2] emphasises that probability gives plausibilities but not actions.

Let us distinguish between the world state space X and a hypothesis space Θ indexing possible models of X, and A the action space,

First of all most literature is touching on loss that is defined as:

Definition 4.2 (Loss function). Let Θ be a hypothesis/state space and A an action space. A *loss function* is

$$L : A \times \Theta \to [0, \infty),$$

where $L(a, \theta)$ is the cost of taking action a when the true state is θ.

The limit of loss function is that if we encounter some x where $L(x) = -\infty$ we are unable to recover, since it is infinite. To bridge this gap, we use utility:

Definition 4.3 (Utility function, revisited). Let Θ be a hypothesis/state space and A an action space. A *utility function* is

$$U : A \times \Theta \to [-\infty, \infty),$$

where $U(a, \theta)$ is the value or benefit of taking action a when the true state is θ.

Given a belief (posterior) P on Θ—recall that a belief state is simply a probability distribution encoding our uncertainty about which model θ is correct—the *Expected Utility* of action a is

$$\mathbb{E}[U(a \mid P)] = \int_{\Theta} U(a, \theta)\, P(d\theta).$$

A Bayes (optimal) action is any

$$a^* \in \arg\max_{a \in A} \mathbb{E}[U(a \mid P)].$$

For sequential or planned decisions, replace P by the current posterior and optimise expected cumulative utility under the chosen policy.

Practically, this yields three light prescriptions:

[2]Jaynes, E. T. (2003). *Probability Theory: The Logic of Science.* Cambridge University Press.

- Allocate effort proportional to potential utility variance: spend more data and deliberation where the difference in utility between best and worst outcomes is large.
- Prefer actions that preserve option value: choose reversible experiments so that future utility can be maximised as new information arrives.
- Encode non–negotiable constraints as negative infinite utility (or arbitrarily low values) so the optimisation strictly avoids those regions.

4.3 Planning as Control Under Uncertainty

We can now write a very compressed equation for "the way of success" in this book:

> *Choose a utility function U, a belief state P, and design a policy π that, when coupled with the world dynamics f, keeps essential variables within R while steering the state into regions of high U, under the constraints of requisite variety.*

In symbols, very schematically:

$$\pi^* = \arg\max_{\pi} \mathbb{E}_P \left[\sum_{t=0}^{T} U(x_t) \right]$$

subject to

$$x_{t+1} = f(x_t, \pi(y_{\leq t}), w_t), \quad E(x_t) \in R, \quad V_R \geq V_D - \varepsilon.$$

Of course, in real life you will never compute this optimisation exactly. The point of writing it down is different:

- it makes explicit *what you are trying to do*;
- it shows which parts can be improved locally (better beliefs, more actions, fewer disturbances);
- it shows where improvement is bounded by hard laws (limited variety, limited attention, limited data).

4.4 Macro-Levers, Micro–Levers and Daily Systems

Micro-levers and Daily Systems

All of this can stay philosophical if you are not careful. To make it concrete, I built what I call *micro–levers*: small, repeatable actions that touch each major pillar of my life every day.[3]

The pattern is:

- for each essential variable or important dimension,
- define a minimal action that improves it slightly (two glasses of water, one page of a serious book, one message to a friend, one line of code or marketing copy);
- design the environment so that doing these is easier than not doing them.

Formally, you are designing a low–complexity policy π with good local properties: it increases the probability that E_t stays in R and that the state slowly drifts toward better regions.

Empirically, this kind of system beats big emotional pushes. When I tried to "change my life" with huge one–off efforts, the dynamics looked like a damped oscillation: strong push, friction, regression. When I switched to micro–levers, I got a slower but monotone drift. On a graph, it looked boring. Inside, it felt like compound interest.

Macro–Levers

While micro-levers handle daily maintenance, some changes require larger structural moves—what I call *macro-levers*. These are infrequent but high-impact decisions that reshape the state space itself:

- changing cities or countries (alters $R(x_0)$ dramatically);
- ending or starting a major relationship;
- switching careers or industries;

[3]Concretely: one health action (exercise, nutrition), one learning action (reading, practice), one relationship action (message, call), one work action (focused output), one recovery action (rest, reflection). These are minimal thresholds, not maximums.

- making a large financial commitment (buying a house, starting a company);
- acquiring a credential that unlocks new action spaces.

Macro-levers are dangerous precisely because they are powerful: they can expand your reachable set or collapse it. The framework suggests: use micro-levers to keep essential variables stable, accumulate information and resources, then deploy macro-levers when you have enough slack to absorb the transition costs and enough data to estimate the new $R(x_0)$ accurately.

4.5 The Limits of Control

It is tempting, especially for mathematically trained people, to believe that with enough modeling we can fully control our success. The reality is harsher:

- there are unknown unknowns: states and transitions we did not even include in our X;
- there are adversarial agents with their own utility functions, sometimes hidden;
- there are structural forces (institutions, markets, ecosystems) that have more variety than we can ever match.

These limitations stem from a mathematical fact: we are neither omniscient nor omnipotent.

Omniscience would require: A belief state P that achieves $P(x_t|y_{0:t}) = \delta_{x_t}$ (perfect knowledge of the current state), plus perfect knowledge of the dynamics f. In finite time with finite data, we cannot achieve this. Even with infinite data, the world may be genuinely stochastic (not merely unknown). Moreover, the state space X itself may be incomplete—there are variables we did not model.

Omnipotence would require: An action repertoire with variety V_R matching every possible disturbance variety V_D. But disturbances come from an environment that evolves independently of us. For complex systems, V_D grows without bound while V_R is biologically and physically finite.

As we have already said, the law of requisite variety formalizes this:

$$V_O \geq V_D - V_R.$$

Since we cannot make V_R arbitrarily large, the best we can do is bound V_O from below. We cannot make it zero, no matter how hard we try.

Remark 4.1 (Humility and strategy). Understanding these limits is not pessimism; it is clarity. Since we cannot control everything, we must choose carefully: (1) which disturbances to defend against (reduce V_D), (2) which skills and tools to develop (increase V_R), and (3) which domains to focus on where the remaining chaos V_O is tolerable.

Chapter 5

Application: Learning an Ultra–Hard Skill

This chapter answers the practical question: *how do we apply the framework to something concrete*? We will run the full machinery on a live example: learning an ultra–hard skill (e.g. becoming research–level in a new mathematical field or mastering a deep technical domain) starting from near zero.

5.1 Define the Model: States, Self, Essential Variables

Let the world X encode everything relevant to the skill and your life. Select a coarse state representation that you can actually track:

- knowledge graph nodes (topics, theorems, tools) mastered or not;
- capability levels on core subskills (proof writing, computation, tool use, experimentation);
- support variables (sleep, time budget, cash buffer, social support).

Define essential variables $E = (E_1, E_2, E_3)$ for survival during the project:

- E_1: health markers (sleep hours, injury risk), with survival region R_1 (e.g. 7–8h sleep, no chronic pain);

- E_2: financial stability (runway months, debt limits), region R_2 (e.g. >6 months runway, debt not increasing);
- E_3: relationships (weekly meaningful contacts), region R_3 (e.g. at least two non–work interactions per week).

The survival region is $R = R_1 \times R_2 \times R_3$. The project is only "successful" if $E(x_t) \in R$ throughout; otherwise learning collapses.

5.2 Reachable Set and Requisite Variety

Estimate the reachable set $R(x_0)$ over a finite horizon (say 12–18 months). List constraints: available hours per week, prerequisite courses, access to mentors or labs, visa or location limits. Identify disturbance variety V_D: randomness in health, work emergencies, resource access.

Design regulator variety V_R to match: multiple study modes (textbook, papers, problem sets, projects), multiple feedback channels (peers, mentors, online forums, code tests), multiple fallback plans (pause, reduce scope, switch subfield). Aim for $V_R \geq V_D - \varepsilon$ so the outcome variety V_O stays tolerable. If V_R is too small, either reduce project scope (lower V_D) or add tools and allies (raise V_R).

5.3 Policy Design: Micro–Levers for a Hard Skill

Construct a low–complexity policy π that you can execute daily with minimal willpower. Example micro– levers:

- **Core practice:** 90 minutes of high– difficulty problems or derivations, timed, with written solutions. Observable: problems solved per week, error rate.
- **Concept map:** update a knowledge graph after each study block; observable: nodes added, links clarified.
- **Feedback loop:** weekly review with a mentor or peer; observable: misconceptions corrected, next focus area chosen.
- **Retrieval:** spaced repetition on definitions and lemmas; observable: retention rate.

- **Project anchor:** one small end—to—end project each month (e.g. rederive a key result, reproduce an experiment); observable: project completion and replication fidelity.
- **Support levers:** fixed sleep window, two social contacts per week, weekly money check—in to keep E inside R.

These micro—levers instantiate the law *stability before optimisation*: they keep E stable while π pushes the state toward competence.

5.4 Policy and Bayesian Learning Loop

Maintain a belief state P_t over your own capability parameters (speed of learning, error rates, prerequisite gaps). After each week, update P_t with data: problem success fraction, recall scores, mentor feedback. Then choose actions that maximise expected utility for the coming week: which subtopic to attack, whether to deepen proofs or shift to applications, whether to schedule an extra review.

In symbols, with a short horizon T:

$$\pi_t = \arg\max_{a \in \text{plans}} \mathbb{E}_{P_t}\Big[U(x_{t+1})\Big],$$

subject to $E(x_{t+1}) \in R$ and practical limits (hours, money). The utility U can weight mastery (problem success), momentum (streaks maintained), and integrity (no violation of R).

5.5 Failure Modes and Replans

Common breakdowns and responses:

- **Overload**: E_1 (health) leaves R_1. Action: cut load, switch to lighter tasks (reading, flashcards), increase sleep.
- **Resource shock**: E_2 (runway) leaves R_2. Action: pause scope, find income buffer, replan horizon.
- **Isolation**: E_3 drops. Action: schedule standing peer sessions; join a forum; ask one mentor question per week.

- **Stall in comprehension**: posterior P_t shows low progress. Action: add tutor/mentor (increase V_R), change study modality, decompose into smaller subskills, or pick a simpler subdomain (reduce V_D).

Each replanning step is an ultrastable move: change the policy structure rather than pushing harder on a failing policy.

5.6 Complementary Completion and Transfer

Define a success predicate for the project: $S(x) = 1$ if you can solve a representative set of hard problems unassisted, explain the core results, and reproduce at least one published result or benchmark. When $S = 1$, shift the utility to transfer: apply the skill to a real domain (research, product, teaching). Keep a maintenance policy with low V_R (light practice and spaced repetition) so E stays inside R while the skill compounds in other systems of your life.

Afterword: Synthesis

The scope of this framework:

This book claims that the four structures—states, possibility spaces, dynamics, and utility—are necessary for coherent thinking about success. It does not claim they are sufficient. The framework shows *which* problems are tractable and which are not, but it does not remove luck, structural injustice, or the fundamental bounds on what any finite system can achieve.

The honest assertion is: if you explicitly define your state space, identify your essential variables, map your reachable set, respect variety constraints, design stabilizing feedback, maintain coherent beliefs, and optimize expected utility subject to survival constraints, then your expected trajectory will be *demonstrably better* than trajectories from systems that ignore these structures.

What follows from here is engineering: apply these principles to the domains where you choose to act, accept the limits imposed by the laws of requisite variety and information, and build routines that embed the framework so deeply that it runs without constant conscious effort.

The rest is yours.

Leave a Review!

If you found this book helpful, I would be incredibly grateful if you could take a moment to leave a review on the platform where you purchased or accessed it. Your feedback not only helps me improve future editions but also assists other readers in discovering the book. You can leave Amazon review here: `https://www.amazon.com/dp/B0FHG3SBV5/`

About the Author

Mikołaj Mocek is a Co-Founder, CEO, and CTO at iminti, a company focused on creating responsible artificial intelligence and innovative software solutions. He is also the author of this book, *Mathematica Successūs: A Formal Approach on Success, Systems and Self*, which applies mathematical rigor to the principles of personal and professional achievement.

With over six years of experience as a senior software engineer, Mikołaj specializes in developing scalable web applications, network infrastructures, and mobile apps. His expertise spans full-stack development, data analysis, and machine learning, where he designs custom solutions for training and evaluating both proprietary and external models. Passionate about the mathematical foundations of AI, including neural networks, he combines engineering precision with entrepreneurial vision.

Beyond technology, Mikołaj is committed to social impact, having developed apps that enhance city accessibility for people with disabilities and reached thousands through iminti Go!.

Mikołaj values continuous learning and interdisciplinary approaches. He believes in leveraging technology and a positive mindset to drive meaningful change, bridging the gap between theoretical mathematics and real-world applications.

Contact

Comments, remarks, suggestions? For this and inquiries, collaborations, or to learn more about Mikołaj's work, you can reach him at:

- Email: `mikolaj.mocek@icloud.com`
- LinkedIn: `https://www.linkedin.com/in/mikolajmocek/`

He would love hearing from you all!

Bibliography

[1] Ashby, W. R. (1956). *An Introduction to Cybernetics.* London: Chapman & Hall.

[2] Åström, K. J., & Murray, R. M. (2008). *Feedback Systems: An Introduction for Scientists and Engineers.* Princeton: Princeton University Press.

[3] Enderton, H. B. (1977). *Elements of Set Theory.* New York: Academic Press.

[4] Folland, G. B. (1999). *Real Analysis: Modern Techniques and Their Applications* (2nd ed.). New York: John Wiley & Sons.

[5] Jaynes, E. T. (2003). *Probability Theory: The Logic of Science.* Cambridge: Cambridge University Press.

[6] Thomas Aquinas. (1920). *Summa Theologica, Part II-II (Secunda Secundae) / Translated by Fathers of the English Dominican Province.* New York: Benziger Brothers.

www.ingramcontent.com/pod-product-compliance
Lightning Source LLC
Chambersburg PA
CBHW072244240926
51846CB00012B/615